Industrial pollution

Dr. Hemant Pathak

ISBN: 149236116X
ISBN-13: 978-1492361169

DEDICATION

Dedicated to Shri Sainath Maharaj the all omnipotent of world the most merciful.

CONTENTS

Foreword

Industrial pollution; provides a unique insight into the problems our planet faces in terms of clean environment, and what to do about it. This is the only books Written for academics, researchers and practitioners working in environmental pollution and management field, expressed comprehensive and interdisciplinary focus on the ecological issues associated with industrial pollution to provide a complete picture of current environmental problem from cause to effect to solution

This book made of 10 years consistently research on environmental issues, makes it ideal source for students, teachers, industrialist, environmental experts and economists.

This book provides an essential guide to researchers, it offers: various causes of pollution; on the challenges and experiences in present scenario.

Simply explained, Industrial pollution is an important book bringing together diverse viewpoints from academia and environmental agencies and regulators, for all who wish to make a difference in how to plan and manage our Environmental resources.

Dr. Hemant Pathak

M.Sc. (Gold medalist), Ph. D.

Assistant Professor of Engineering Chemistry

Indira Gandhi Govt. Engineering College,

Sagar, MP, India

Glossary	
Abatement	The reduction or elimination of pollution.
Acid rain	The precipitation of dilute solutions of strong mineral acids, formed by the mixing in the atmosphere of various industrial pollutants
Act	A law
Acute Exposure	One or a series of short-term exposures generally lasting less than 24 hours.
Aerosol	Particles of solid or liquid matter than can remain suspended in air from a few minutes to many months depending on the particle size and weight.
Air pollution	Toxic or radioactive gases or particulate matter introduced into the atmosphere, usually as a result of human activity.
Air Toxic	Any air pollutant for which a ambient air quality standard does not exist that may reasonably be anticipated to cause cancer, developmental effects, reproductive dysfunctions, neurological disorders, heritable gene mutations or other serious or irreversible chronic or acute health effects in humans.
Ash	Incombustible residue left over after incineration or other thermal processes.
Atmosphere	The 500 km thick layer of air surrounding the earth which supports the existence of all flora and fauna.
Biodiversity	A large number and wide range of species of animals, plants, fungi, and microorganisms. Ecologically, wide biodiversity is conducive to the development of all species.
Climate change	A regional change in temperature and weather patterns. Current science indicates a discernible link between climate change over the last century and human activity, specifically the burning of fossil fuels.
Combustion	Burning. Many important pollutants, such as sulfur dioxide, nitrogen oxides, and particulates (PM-10) are combustion products, often products of the burning of fuels such as coal, oil, gas, and wood.

Contamination	The act of polluting or making impure; any indication of chemical, sediment, or biological impurities.
Dust	Solid particulate matter that can become airborne.
Ecosystem	An interactive system that includes the organisms of a natural community association together with their abiotic physical, chemical, and geochemical environment.
Effluent	Municipal sewage or industrial liquid waste (untreated, partially treated, or completely treated) that flows out of a treatment plant, septic system, pipe, etc.
Emission	Release of pollutants into the air from a source. We say sources emit pollutants. Continuous emission monitoring systems (CEMS) are machines, which some large sources are required to install, to make continuous measurements of pollutant release.
Exposure	The concentration of the pollutant in the air multiplied by the population exposed to that concentration over a specified time period.
Fossil fuels	Fuels such as coal, oil, and natural gas; so-called because they are the remains of ancient plant and animal life.
Global warming	increase in the average temperature of the earth's surface.
Greenhouse gases	Atmospheric gases such as carbon dioxide, methane, chlorofluorocarbons, nitrous oxide, ozone, and water vapor that slow the passage of re-radiated heat through the Earth's atmosphere.
Hydrocarbons	Compounds containing various combinations of hydrogen and carbon atoms. They may be emitted into the air by natural sources (e.g., trees) and as a result of fossil and vegetative fuel combustion, fuel volatilization, and solvent use. Hydrocarbons are a major contributor to smog.
Industrialized	Nations whose economies are based on industrial production and the

countries	conversion of raw materials into products and services, mainly with the use of machinery and artificial energy (fossil fuels and nuclear fission).
Kyoto Protocol	An international agreement adopted in December 1997 in Kyoto, Japan. The Protocol sets binding emission targets for developed countries that would reduce the emissions on average 5.2 percent below 1990 levels.
Micro- (μ)	The metric prefix for one millionth of the unit that follows.
Microgram (μg)	One millionth of a gram: 1 μg = 10^{-6} g = 0.001 mg.
Mitigation	Actions taken to avoid, reduce, or compensate for the effects of environmental damage. Among the broad spectrum of possible actions are those that restore, enhance, create, or replace damaged ecosystems.
Monitoring	Periodic or continuous surveillance or testing to determine the level of compliance with statutory requirements and/or pollutant levels in various media or in humans, plants, and animals.
Non-point pollution	diffuse pollution mainly from agriculture or dumping grounds. It is difficult to collect for treatment.
Opacity	The amount of light obscured by particle pollution in the atmosphere. Opacity is used as an indicator of changes in performance of particulate control systems.
Ozone	A gas which is a variety of oxygen. High concentrations of ozone gas are found in a layer of the atmosphere - the stratosphere - high above the Earth. Stratospheric ozone shields the Earth against harmful rays from the sun, particularly ultraviolet B.
Ozone depletion	the reduction of the protective layer of ozone in the upper atmosphere by chemical pollution.
PM-10	particulate matter less than 10 microns in diameter.

Particulate pollution	pollution made up of small liquid or solid particles suspended in the atmosphere or water supply.
ppb/ ppm	Units commonly used to express contamination ratios, as in establishing the maximum permissible amount of contaminant in water, land, or air.
Plume	A visible or measurable discharge of a contaminant from a given point of origin that can be measured according to the Ringelmann scale.
Pollutants (pollution)	Unwanted chemicals or other materials found in the air. Pollutants can harm health, the environment and property. Many air pollutants occur as gases or vapors, but some are very tiny solid particles: dust, smoke, or soot.
Point pollution	polluted water from a defined point. It can be collected as industrial or municipal wastewater and treated by what is often called end-of-pipe technology (environmental technology).
Pollution control	The addition of processes, practices, materials, products or energy to waste streams to reduce the risk posed by pollutants and waste before their release to the environment.
Pollution prevention	The use of processes, practices, materials, products, substances or energy that avoid or minimize the creation of pollutants and waste, and reduce the overall risk to human health or the environment
Public health	the health or physical well-being of a whole community.
Reuse	The reemployment of products or materials, in their original form or in new applications, with refurbishing to original or new specifications as required.
Risk assessment	Methods used to quantify risks to human health and the environment.
Run-off	Precipitation that the ground does not absorb and that ultimately reaches rivers, lakes or oceans.

Smog	A mixture of pollutants, principally ground-level ozone, produced by chemical reactions in the air involving smog-forming chemicals.
Solid waste	non-liquid, non gaseous category of waste from non-toxic household and commercial sources.
Threatened species	species of flora or fauna likely to become endangered within the foreseeable future.
Toxic emissions	poisonous chemicals discharged to air, water, or land.
Toxic waste	garbage or waste that can injure, poison, or harm living things, and is sometimes life-threatening.
Visibility	A measurement of the ability to see and identify objects at different distances. Visibility reduction from air pollution is often due to the presence of sulfur and nitrogen oxides, as well as particulate matter.
Waste	Garbage, trash.
Waterborne contaminants	Unhealthy chemicals, microorganisms (like bacteria) or radiation, found in tap water.
Water quality	The level of purity of water; the safety or purity of drinking water.

1. Introduction

India is the secondly largest and densely populated countries in the world. It is also one of the developing country. The 70% population remains largely rural with only around 30 percent living in urban areas. Rural livelihoods are dominated by agricultural activity. we have the dire necessity to increase fertilizer production for maintaining food security. The process involved in the production of fertilisers generates effluents and the emissions contribute to industrial pollution such as green house effect, stratospheric ozone depletion, acid rain and acidification, eutrophication, soil degradation, technological hazards, chemical mists etc .. with potential damage to human race.

Environmental degradation affects in Indian context more fundamentally, than it does the developed world. It is universally recognized as developing country. economic development is an essential process to erase poverty and hunger. it is also important to protect the environment from pollution at regional as well as national and global levels.

While on the other hand the consequential environmental issues are to be tackled effectively to avoid damages.

The industrial sector is one of the most dynamic sectors of the economy and plays an essential role in economic development and the alleviation of poverty.
In India Industrialization contributes goods, services and jobs, but it is also responsible for pollution and waste.

Pollution is known to harm health and the estimated social costs due to pollution are very large. Indian economies with low levels of industrialization are gradually shifting their dependence from agriculture to the industrial sector, while developed economies, with a high level of industrialization, are shifting from the industrial to the service sector.

But Industrialization has also brought with it a range of problems. Rapid industrial growth lifts people out of poverty but also leads to increased environmental pollution. The industries tend to be clustered together and are highly polluting. Tight regulation using traditional models could therefore do real economic harm.

As a consequence of their rapid and largely unregulated development, many aquatic ecosystems are now under threat and with them the livelihood systems of local people, these problems are broadly recognized, environmental regulation in India must strike a delicate balance between pollution and development.

Many unwanted substances and losses generated by industrial activities, including emissions to air or surface waters and the substances sent to sewage treatment plants, deposited in landfills, released or applied to the land, treated, injected underground, controlled through storage, recycled or burned for energy recovery. economic activities that create pollution from transport to industry and electricity production are themselves important for growth.

The recent monitoring report received from the Central Pollution Control Board (CPCB) indicates that of the 851 industries along the major rivers and lakes in the country which have found discharging their untreated or partially treated effluents into fresh water bodies, 608 units were found to comply with prescribed standards and another 238 industrial units have been closed.

In order to address the issue of Indian scenario a World Bank assisted programme for capacity building was launched in 1997. As on September 30, 2001, out of the total 1,551 large and medium industries identified in 1992 in the 17 categories of highly polluting industries, 1,349 industries have installed the requisite pollution control facilities to comply with the prescribed environmental standards.

177 industries have closed down and only 25 industries have yet to install the necessary pollution control facilities (status as on September 30, 2001). Legal action has been taken against all the defaulting industries.

Only 5 units have not installed the requisite pollution control facilities; legal action has been initiated for all the defaulting units.

2. Classification of Industrial pollution

Intensive Industrialization in India, particularly in northern India has led to rapid depletion of natural resources in recent years.

Industrial pollution and waste can be classified into six categories:

I. toxic chemicals   II. air pollutants

III. greenhouse gases

IV. hazardous and nonhazardous wastes

V. radioactive wastes.

I. Toxic Chemicals

Toxic Chemicals are hazardous to human health and the environment, released from industry include heavy metals, cyanides and pesticides, and can be emitted into air, water or in solid wastes. Every year Indian industries generated million tonnes of toxic chemicals as production related pollutants and waste.

Bio-magnification describes situations where toxins (such as heavy metals) may pass through trophic levels, becoming exponentially more concentrated in the process. However, their impacts are found in polluted groundwater, surface water, and urban and peri-urban refuse dumps. The effects of certain toxic chemicals on the health and development of children and other vulnerable groups are a special concern.

II. Air pollutants

These substances, which include nitrogen oxides, sulfur oxides, carbon monoxide, Total Suspended Particulates (TSP), CO_2 and hydrocarbons (such as methane) and volatile organic compounds, are associated with environmental effects such as smog, acid rain and regional haze, and health effects such as respiratory illness.

Nitrogen oxides are removed from the air by rain and fertilize land which can change the species composition of ecosystems. SO_2 is formed when sulphur-containing fuels like coal and oil are burned. In the same vein, NO_2 and greenhouse gases emissions are attributed to fuel combustion for energy generation in motor vehicles, power stations and furnaces.

Industrial air pollution is primarily derived from energy use. Industry consumes over 40 per cent of commercial energy in India and rest south Asia. These pollutants are emitted from a variety of sources, including residential fuel combustion, motor vehicles and agricultural activities. Industrial sources are also major contributors among them, electric utilities, primary metal smelters and cement kilns. Smog and haze can reduce the amount of sunlight received by plants to carry out photosynthesis and leads to the production of tropospheric ozone which damages plants.

Energy efficiency, therefore, is of utmost importance and it needs least cost investments that industrial firms can easily make to reduce air pollution.

Although emissions of air contaminants are trending downward, reductions from sources such as motor vehicles have been partially offset by increases from certain oil and gas industry subsectors attributed to expanded production. Sulfur dioxide and nitrogen oxides can cause acid rain which lowers the pH value of soil.

Energy efficient technologies are implicit in most investment in 'clean technologies', which reduce pollution through reduced inputs and lower pollution intensities.

III. Greenhouse Gases

These gases, which include carbon dioxide (CO_2), methane and nitrous oxide, are linked to global climate change. Industrial energy use is a major source of CO_2 emissions in India. Carbon dioxide emissions cause ocean acidification, the ongoing decrease in the pH of the Earth's oceans as CO_2 becomes dissolved. The emission of greenhouse gases leads to global warming which affects ecosystems in many ways.

coal mining and the use of coal, greenhouse gases, acid rain and ground level ozone, issues which can be local, regional and global in their impacts.

India has a commitment under the United Nations Framework Convention on Climate Change (UNFCCC) to monitor its complete greenhouse gases emissions. The greenhouse gases emissions from the industry sector amounted to 405.86 million tons of CO_2, 0.15 million tons of CH_4 and 0.21 million tons of NO_2, which amounted to 412.55 million tons of CO_2 equivalent.

IV. Hazardous and Nonhazardous Wastes

Hazardous wastes are industrial waste streams that may contain more than a single chemical or substance. They are typically defined by characteristics such as ignitability, reactivity, corrosivity and toxicity.

Nonhazardous industrial wastes include coal ash, foundry sands, cement kiln dust, mining and mineral processing wastes, oil and gas production wastes, and other wastes that lack the characteristics of hazardous waste.

In India rapid growth of industries has resulted in generation of increasing volume of hazardous wastes. Both indigenously generated and imported from other countries for recycling or reprocessing need scientific treatment and disposal.

For disposal of hazardous wastes few secured landfill sites are available in the country. An illegal dumping of hazardous wastes by the industries may cause severe environmental pollution. The Ministry of Environment and Forests has promulgated Hazardous Wastes (Management & Handling) Rules, 1989 for proper management and handling of hazardous wastes. These rules also deal with the ban for importing a few categories of hazardous wastes.

India has also ratified the Basel Convention on transboundary movement of hazardous wastes in 1992, for controlling and monitoring of import and export of hazardous wastes and its proper management.

v.   Radioactive Wastes

Radioactive wastes are by-products of certain industrial activities, in particular electricity generation.

The processes involved in these industries - which also include other sectors such as electricity generation, petroleum refining, mining, paper production, and leather tanning - are resource-intensive, and tend to produce a disproportionately large amount of hazardous and toxic wastes.

3.   Environmental Impact of Industrial pollution

I.   Change of Climate (including weather events, variable climates that affect food and water supplies and ecosystem changes)

Industrial pollution and waste contribute to climate change as well. The problem is not exclusive to India, but it is perhaps most critical in the south Asia, most densely populated region. Public health, food security, industrial growth, and ecosystems all are currently at greater risk than ever before. More public investment will be needed to manage the growing demand for drinking, industrial and irrigation water supplies.

The anaerobic decomposition of wastes in landfills produces methane, a potent greenhouse gas, and waste incineration releases carbon dioxide. The transportation of wastes to recycling, treatment and disposal sites produces transportation related carbon emissions.

Finally, the materials disposed of as waste must be replaced by more raw materials, which implies further consumption of fossil fuels and additional carbon releases.

II. Health Hazard

The pollution and waste problems regulated by central pollution control board in India are also identified as raising concerns about human health or the environment. Every year people die in India due to respiratory illnesses triggered by air pollution. Globally, scientific research is illuminating the clear linkages between not only air pollution and respiratory complications, but also numerous diseases.

Industrial pollution influences human health and disease. It destroyed natural environment like air, water and soil, and also all the physical, chemical, biological and social features of our surroundings. Consequently alteration to the natural environment, such as air pollution, are also affect to the man-made environment and influences on health.

In one study in the *Journal of Toxicology and Environmental Health*, researchers found that in the developing nations of Asia two-thirds of their health problems are due to urban air pollution.

III. Economic Costs

Industrial pollution have potential effects on humans and the environment, wastes represent inefficiency in industrial production. Wastes impose costs on facilities; consequently waste management increased production cost.

IV. Waste Management

There is numerous types of wastes. But Industrial waste management is major issues. Municipal waste incineration, medical waste incineration, burning of hazardous wastes in cement kilns and backyard waste burning were among the top sources of dioxins. Decisions on how to manage wastes have environmental implications.,

V. Depletion of Natural Resources

Unfortunately polluted water and air are common throughout the world. Inefficient use of materials and energy affects the use of natural resources. India faces a two-fold issue in water security: decreasing water availability and increasing water pollution in groundwater and surface resources. An increasing population and greater industrial activity is putting pressure on present water sources. Per capita water availability has reduced significantly from 1,816 cubic meters per capita in 2001 to 1,588 cubic metres in 2010.

Depletion of natural resources is mediated by the renewability of the inputs used and the degree of recycling undertaken within or among industrial sectors.

4. Environmental Impact of Large Industries

At the time of establishing industrial units, the pollution problems in the industrial operations were given very little attention. These industries were set up with practically no pollution control measures and hence have led to the environmental degradation beyond permissible limits. Therefore, a stage has been reached where adequate and effective pollution control measures are required to be adopted in such industries so that the adverse effects, to the environment, are minimized.

Industries such as cement, glass, ceramic, iron and steel, paper and pulp, refineries, etc, exercise a wide range of environmental impacts. They emit large amounts of nitrogen, sulphur and carbon oxides into the air.

Emissions of lead, arsenic and chromium, both from glass and iron and steel industries, are extremely toxic. Waste disposal from such industries causes extensive water and soil contamination too. Extraction of raw materials causes large-scale surface disturbance and erosion.

5. Water pollution and Industrial waste water

The water we drink are essential ingredients for our wellbeing and a healthy life. Globally, it is recognized as an important natural resource with great economic value. The WHO states that one sixth of the world's population, approximately 1.1 billion people do not have access to safe water and 2.4 billion lack basic sanitation.

Water pollution comes from three main sources: domestic sewage, industrial effluents and run-off from activities such as agriculture. Wastewater from industrial activities is often contaminated with highly toxic organic and inorganic substances, some of which are persistent pollutants and remain in the environment for many years. As per 2009 figures, wastewater generation from industry has been estimated to be 55,000 million m3 per day, of which 68.5 million m3 are dumped directly into local rivers and streams without prior treatment. The Indian government has given provision for the establishment of Common Effluent Treatment Plants (CETP) during the 11th Year Plan which has been effectively piloted with 142 CETPs set up with government assistance.

A safe and sustainable water supply is essential for improving public health, and achieving economic growth and food security in the south Asia like India. Our water resources are facing degradation due to a range of problems, such as overexploitation, mismanagement, and natural and anthropogenic contamination.

This water pollution affects the health and quality of soils and vegetation. Some water pollution effects are recognized immediately, whereas others don't show up for months or years. Estimation indicates that more than fifty countries of the world with an area of twenty million hectares area are treated with polluted or partially treated polluted water including parts of all continents and this poor quality water causes health hazard and death of human being, aquatic life and also disturbs the production of different crops.

In fact, the effects of water pollution are said to be the leading cause of death for humans across the globe, moreover, water pollution affects our oceans, lakes, rivers, and drinking water, making it a widespread and global concern.

In present scenario due to industrialization and increased population, carry the industrial and municipal effluents that are ultimately carried that polluted water to the canals and rivers.

The untreated Industrial production and natural resource use and conversion often result in the creation of large amounts of water-borne pollutants. The major categories of water-borne pollutants are oxygen demand substances, measured by biological oxygen demand (BOD5) and chemical oxygen demand (CODs); other standard industrial pollutants (such as total suspended solids), ammonia, phosphorous, sulphide, nitrate, sulphate, chloride, oil and grease; and polluting characteristics (such as pH.)

Some industrial facilities generate ordinary domestic sewage that can be treated by municipal facilities. Industries that generate waste water with high concentrations of conventional pollutants (e.g. oil and grease), toxic pollutants (e.g. heavy metals, volatile organic compounds) or other non conventional pollutants such as ammonia, need specialized treatment systems. Some of these facilities can install a pre-treatment system to remove the toxic components, and then send the partially-treated waste water to the municipal system. Industries generating large volumes of waste water typically operate their own complete on-site treatment systems.

6. Bhopal Disaster: A case study in context to India

India has experienced rapid industrial growth since the enactment of the economic liberalization policies in 1991. Economic liberalization has accounted for a substantial impact on the manufacturing industry through an increase in the presence of manufacturing units, from 98,379 in a pre-liberalization period of 1987 to 1,40,355 industrial units in 2007 reflecting a 42.67% growth during this 20 year period, and a rise in the production capacity and output within individual manufacturing facilities.

The Bhopal disaster was a major leak of toxic chemical gases incident in India, considered the world's worst industrial disaster. It occurred on the night of 2–3 December 1984 at the Union Carbide India Limited (UCIL) pesticide plant in Bhopal, Madhya Pradesh. Over 500,000 people were exposed to methyl isocyanate gas and other chemicals. The toxic substance made its way in and around the shanty towns located near the plant. The official immediate death toll was 2,259. The government of Madhya Pradesh confirmed a total of 3,787 deaths related to the gas release and an estimated 1,00,000 persons continue to suffer from chronic eye, respiratory, gastrointestinal and psychiatric illnesses and cannot live or work normally. The environment and economy in Bhopal has also suffered due to water, soil, food and livestock contamination. The poorest population living in the slums around the Union Carbide plant suffered the brunt of these impacts.

The Bhopal disaster underlines the problem governments confront in formulating a response to disaster situations when poverty levels are high and health infrastructures and government resources are severely limited. This brings into focus the need for private multinational (or public) industries to take some responsibility towards the environments and populations they are located in. The chemical process employed in the Bhopal plant had methylamine reacting with phosgene to form MIC, which was then reacted with 1-naphthol to form the final product, carbaryl. Factors leading to the magnitude of the gas leak mainly included problems such as; storing MIC in large tanks and filling beyond recommended levels, poor maintenance after the plant ceased MIC production at the end of 1984, failure of several safety systems due to poor maintenance, and safety systems.

The acute symptoms were burning in the respiratory tract and eyes, blepharospasm, breathlessness, stomach pains and vomiting. The causes of deaths were choking, reflexogenic circulatory collapse and pulmonary oedema.

In 1982 tubewells in the vicinity of the UCIL factory had to be abandoned and tests in 1989 performed by UCC's laboratory revealed that soil and water samples collected from near the factory and inside the plant were toxic to fish. Several other studies had also shown polluted soil and groundwater in the area.

7. Control of Industrial pollution

Pollution control is a big problem associated with developing countries like India. All anti-pollution laws and measures, become ineffective, in the absence of proper monitoring system. This can be accomplished through regular monitoring by a competent government agency. It is essential to control of emissions and effluents into air, water or soil under environmental management.

Industrial sewage is typically treated by centralized plants are designed to control BOD, COD and suspended solids. Without pollution control, the waste products from consumption, agriculture, heating, mining, manufacturing, transportation and other human activities, whether they accumulate or disperse, will degrade the environment. pollution prevention and waste minimization are also necessary for pollution control .

Well-designed and operated systems (biological treatment) can remove 85 percent or more of these pollutants. Most municipal plants are not designed to treat toxic pollutants found in industrial waste water.

8. Prevention of Industrial pollution

Natural resources have to be prolonged to their completely use to maintain the aim for continual economic growth and lessen environmental impacts. This involves reducing wastage in operations, utilizing waste products through recycling and recovery practices to further ensure the long-term availability and usefulness of natural resources. Prevention of Industrial pollution can be achieved by reduction or elimination of wastes and pollutants at their sources. For all the pollution that is avoided in the first place, there is that much less pollution to manage, treat, dispose of, or clean up. This goal may be achieved by some activities such as:

I. Implementing better housekeeping practices to minimize leaks and fugitive releases from manufacturing processes

II. Redesigning products to cause less waste or pollution during manufacture, use, or disposal altering production processes to minimize the use of toxic chemicals

III. To reduce energy consumption Pollution prevention within industry generally receives the most attention.

IV. Planting pest resistant crops can reduce or eliminate the need for chemical pesticides, thereby reducing the water, air, and soil pollution that results from the manufacture and use of agricultural chemicals.

V. In office settings, simple steps such as making double sided copies.printing drafts on the back sides of discarded paper can substantially reduce the consumption and disposal of paper products.

VI. To minimizing the use of toxic household chemicals such as drain cleaners and herbicides will reduce the amount of hazardous chemicals that eventually end up in the environment.

VII. Keeping the environment clean and managing the wastes with the Guide lines of respective Government. The Exhausts from the Automobiles and workshop machinery should be controlled.Repair and replacement of leaking and malfunctioning equipment

VIII. The ISO standards must be followed strictly for Industrial usage.

IX. To used Eco-friendly means like bicycle, bike etc. Must used public transportation means like bus for routine jobs. Administration must promoted car pool to office and back.

X. Reduce the use of aerosols in the household. Promote the afforesting. Switch-off all the lights and fans when not required. Promoted to sharing of room with others when the air conditioner, cooler or fan is on.

In the last, India's higher economic growth is increased consumption of the natural resources and increased waste generation that contributes to ecological degradation, which is estimated at around 5% of India's Gross Domestic Product (GDP). Some of the key areas of waste generation are liquid waste, Industrial waste including hazardous wastes, municipal wastes and e waste.

Environmental management assumes paramount importance in this perspective to address the numerous issues relating to pollution control, safety etc., and to minimize the degradation of the environment on account of developmental activities.

9. References

1. U.S. Environmental Protection Agency, Office of Pollution Prevention. Pollution Prevention 1991: Progress on Reducing Industrial Pollutants. EPA 21P-3003. Washington: U.S. EPA, October 1991, pp. 6–7.

2. Harry Freeman et al. "Industrial Pollution Prevention: A Critical Review." Journal of Air and Waste Management 42, no. 5 (May 1992): 619–620.

3. U.S. EPA, Pollution Prevention 1991, pp. 6–7.

4. U.N. Commission on Environment and Development, Our Common Future. United Nations, 1987.

5. Central Pollution Control Board (2012). National Summary, available at http://www.cpcb.nic.in/FinalNationalSummary.pdf.

7. Chen, Y., A. Ebenstein, M. Greenstone and H. Li (2011). The Long-Run Impact of Air Pollution on Life Expectancy: Evidence from China's Huai River Policy," Mimeo, MIT.

8. Center for Science and the Environment (2012) About CSE, available at http://www.cseindia.org/node/214

9. Currie, J., E.A. Hanushek, E. M. Kahn, M. Neidell and S.G. Rivkin (2009). Does Pollution Increase School Absences? The Review of Economics and Statistics, 91(4) 682-694.

10. Dasgupta, N. (2000). Environmental Enforcement and Small Industries in India: Reworking the Problem in the Poverty Context. World Development, 28(5) 945-967.

11. Dasgupta, S., B. Laplante and N. Mamingi (2001). Pollution and Capital Markets in Developing Countries. Journal of Environmental Economics and Management, 42(3) 310-335.

12. Stavins, R. (2003). Experience with Market-based Environmental Policy Instruments. Handbook of Environmental Economics I, 355-435.

13. Tietenberg, T. (1998). Disclosure strategies for pollution control. Environmental and Resource Economics 11, 587–602.

14. Uchida, E., S. Rozelle and J. Xu (2009). Conservation payments, liquidity constraints, and offfarm labor: Impact of the Grain-for-Green program on rural households in China. American Journal of Agricultural Economics, 91(1), 70–86.

15. Wang, H., J. Bi, D. Wheeler, J. Wang, D. Cao, G. Lu, and Y. Wang (2004). Environmental Performance Rating and Disclosure: China's GreenWatch program. Journal of Environmental Management, 71(2), 123–33.

16. World Bank (2010). World Development Report: Development and Climate Change, available at http://wdronline.worldbank.org//worldbank/a/c.html/world_development_report_ 2010/abstract/WB.978-0-8213-7987-5.abstract.

17. Water Pollution Effects, (2006). In Grinning Planet, Saving the Planet One Joke at a Time. Retrieved from http://www.grinningplanet.com/2006/12-05/water-pollution-effects.htm

18. Woodruff, T. J., Parker, J. D. & Schoendorf, K. C. (2006). Fine Particulate Matter (PM2.5) Air Pollution and Selected Causes of Post neonatal Infant Mortality in California, Environmental Health Perspectives. 114(5), pp. 786–790.

19. World Bank, (2002). What Do We Know About Air Pollution?—India Case Study, Urban Air Pollution, South Asia Urban Air Quality Management Briefing Note No. 4, pp. 1-4.

20. World Health Organization (WHO), (2010a). Air Quality: Volcanic Ash Cloud over Europe. Retrieved from http://www.euro.who.int/en/what-we-do/health-topics/environmental-health/air- quality/volcanic-ash-cloud-over-europe

21. World Health Organization (WHO), (2010b). The World Health Report - Health Systems Financing: The Path to Universal Coverage. http://www.who.int/entity/whr/2010/whr10_en.pdf

ABOUT THE AUTHOR

Dr. Hemant Pathak held positions as Assistant Professor in the department of chemistry, Govt. Indira Gandhi Engineering College, Sagar, MP, India. He had extensive experience in teaching, research and administrative management.

Dr. Pathak received his Ph.D. degree in chemistry from Dr. Hari Singh Gour Central University, Sagar, India and M.Sc. Gold medalist from Jiwaji University, Gwalior. He has published 10 books and more than 50 research papers in reputed International and National journals and received several awards. He is a member of editorial boards and reviewer boards of several international journals and societies. His area of specialization includes Engineering Chemistry and Environmental Pollution management.